Coastal Bears

Keith Scott

Thanks

K. Scott

aug. 13

2019

hancock
house

ISBN 0-88839-626-0
Copyright © 2006 Keith Scott

Cataloging in Publication Data

Scott, Keith Vincent, 1936-
    Coastal bears / Keith Scott.

ISBN 0-88839-626-0

1. Bears—British Columbia—Pacific Coast. 2. Bears—Alaska—Pacific Coast. 3. Bears. I. Title.

QL737.C27S356 2006   599.78'097111   C2006-901529-5

Printed in Indonesia—TK

Editor: Theresa Laviolette
Image editor: Laura Michaels
Production: Mia Hancock
Cover design: Mia Hancock
Photography: Keith Scott unless otherwise credited
Cover Photo: Bill Kirsopp

Published simultaneously in Canada and the United States by

HANCOCK HOUSE PUBLISHERS LTD.
19313 Zero Avenue, Surrey, B.C. Canada V3S 9R9
(604) 538-1114 Fax (604) 538-2262

HANCOCK HOUSE PUBLISHERS
1431 Harrison Avenue, Blaine, WA U.S.A. 98230-5005
(604) 538-1114 Fax (604) 538-2262

Website: www.hancockhouse.com
Email: sales@hancockhouse.com

CONTENTS

Credits and Dedication

Normally when anybody hikes with me they don't do it again because I like to go into very rough areas. Some people also think that I have a few screws loose. This isn't true. Conservationist Jim Kahl has hiked with me many times and I was happy to hear from him that I haven't got a few loose screws; he says they're all loose. I am sure that Jim remains friends with me in order to inherit my camera equipment and hiking gear because he has commented many times on my socks that double as footwear and lens covers.

Jim was kind enough to supply two photographs that he took when he was hiking near McNeil River in Alaska. I appreciated his help on one occasion in assisting me in getting a road closed. Eventually this road was turned into a wilderness-hiking trail in Banff National Park, Alberta, proving that people can accomplish things with little more than conviction.

Conservationist Bill Kirsopp has five of his photographs in this book, three taken on a trip to Brooks Falls, Alaska. I appreciate his conservation efforts and his friendship. Bill thinks that I'm as sharp as a basket-ball. He also took pictures of me when I had no cloths on. He was kinky enough to send me a photo (right).

It's great to have their photos in the book.

To my wife, Frances, and sons, Steven and Donny; I appreciate their company on my hiking trips. I also appreciate the Foreword my son, Donny, wrote for this book.

Also, to the many people that hike with me, like Doug Cluff who puts up with my foolishness in the woods, which is a real feat.

(opposite page, left) Jim Kahl.

(opposite page, bottom) One of my sneakers that Jim Kahl wants to inherit.

(below) Bill Kirsopp and one of his grey owl friends. (Photo: Bill Kirsopp)

5

Foreword

Having Keith Scott as a father has been an exceptional experience. I grew up believing that all families spent most of their time hiking, fishing, and taking pictures of bears. Indeed, one of my earliest memories is of my mother, Frances, my father, and I wading across a freezing cold river deep in the heart of the Yukon. Crossing a river with water up to your waist was what I always thought having fun meant, and I was right. Our family has always enjoyed — and endured — the wilderness. My brother Steven, who has continued the family tradition, started early. When he was only nine he hiked twenty-five kilometers into the heart of bear country.

People must assess their lives at some point, and choose a path that will make them happy. If the path is well chosen, they will better understand their part in the cycle of life. There are many who sit behind a desk most of their lives dreaming about adventure. My father was once one of these people; however, his love for nature compelled him to abandon a safe existence. Since then he has lived a life most people only dream about — or dread. Being curious about bears is far from making it a mainstream hobby but fulfilling this interest is what has made my father happy. This interest has landed him in many precarious situations, but he thrives on the excitement. His life has been well chosen.

I have accompanied my father on many trips and on each one I have discovered something new and wonderful about the relationship between humanity and nature. Once, when hiking in the Rocky Mountains with my father, I recall coming to a lake where careless campers had tossed their garbage into the water. We spent the next few hours wading into the lake and removing the mess. This was the first time I realized what it meant to be concerned about nature.

Another time, at home, I saw a spider crawling across the carpet. The natural reaction might be to squish the crawling nuisance, but my father spent twenty minutes trying to get the spider onto a piece of newspaper then carefully taking it out side. But that wasn't all; he spent the next ten minutes watching the spider to make sure it was okay. This level of concern is representative of my father's relationship with nature, which contrasts greatly with that of most people I have met. Nature is often considered an impediment to progress and is largely taken for granted. Few people think twice about getting rid of a bothersome tree — they think cutting it down will improve the view. The person who throws a chip bag out the window of his gas-guzzler is the same one who buys environmentally friendly detergent the next day.

My father was fighting for the protection of wildlife and wildlife habitat long before it became politically correct — not spewing environmental rhetoric, but by doing something. Jim Kahl, one of my father's good friends, expressed a belief shared by many people who know my father. "Over the years I have come to really trust Keith's knowledge of wildlife and his truly deep sense of commitment towards protecting it. He will commit to action, while others are content to talk."

Each summer our adventures begin again, and I continue to learn new ways of appreciating nature. Although my father leads a curious life, it is both honest and admirable.

— Don Scott

Steven Scott relaxing on hillside.   *Donny Scott hiking.*

Should you ever get a chance to hike with me, don't. I'm a very slow hiker but I move continually. Many of the places I go are very rugged and I'm always looking for bears. I am very successful at finding them and I have been in many awkward situations with these misunderstood animals.

One time a grizzly charged me. With no bear spray, I dropped to the ground. The bear sniffed my body then ran away. This was exciting, to say the least. People that know me say that this is the best picture I have of myself.

I have also been told that I camp in ridiculous places but camping under the glacier was like being in a jewel.

1: Coastal Areas Hiked

Areas Hiked Looking for Bears

Newfoundland
Nova Scotia
New Brunswick
Maine, U.S.A.

Brooks Falls, Alaska
Bill Kirsopp

Hudson Bay

McNeil River, Alaska
Jim Kahl

Kodiak Island, U.S.A.

Admiralty Island, U.S.A

Anan Creek, U.S.A.

Fish Creek, U.S.A

British Columbia, Canada

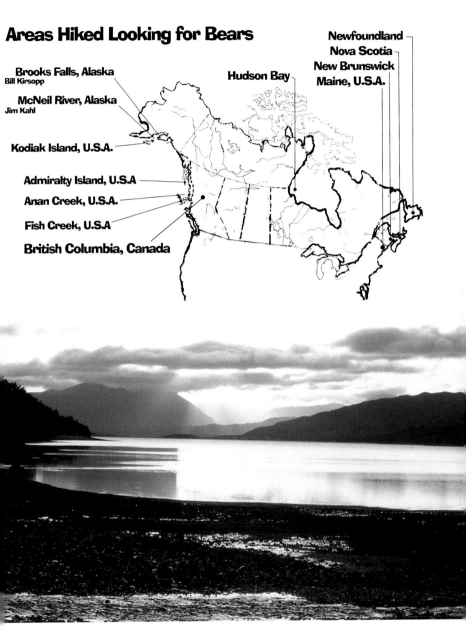

Larsen Bay, where I saw grizzlies, Kodiak Island, Alaska, U.S.A.

9

Great bear viewing areas where other spectacular species also abound.

Gulls at Anan Creek, Anan Creek, Alaska, U.S.A.

Viewing platform built by the United States Forest Service. Fish Creek, Hyder, Alaska, U.S.A.

Portland Canal
The border of British Columbia and Alaska

British Columbia, Canada

Humpback whale singing. Admiralty Island, Alaska, U.S.A.

Fall Colors, State of Maine, U.S.A.

Atlantic Salmon, New Brunswick, Canada. The only salmon on the Eas Coast are the Atlantic salmon that lives to spawn more than once. Als enjoyed by black bears.

*Nova Scotia,
Canada
Fine black
bear habitat.*

*Hudson Bay,
Manitoba,
Canada
Polar bear
habitat.*

13

2: Black Bears

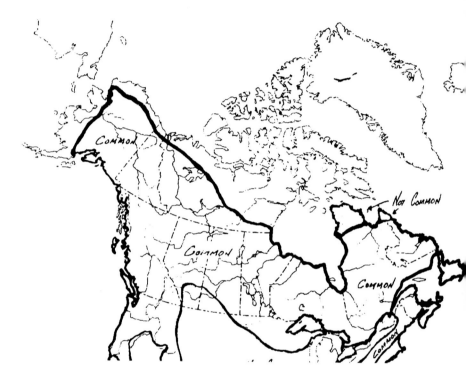

Areas where black bears are commonly seen.

There are eight species of bears in the world. In North America there are three — the American black bear, the grizzly bear, and the polar bear. The most dangerous living creatures in North America are humans and then domestic dogs. I have fought with and been injured by both humans and dogs, but I have never fought, or been injured by, a bear.

Black bears are a massive, bulky animal that can grow up to five feet (one and a half meters) in length. Their normal weight is around 200 pounds (ninety kilograms), however bears eating a very rich diet, such as salmon, can weigh much more.

"I'm going to eat the whole thing."

"Keep up with me."

"This is what I think of you people."

Black bears are usually charcoal black and many of them have a patch of white fur on their chest. They can be light or dark brown, white, or even black with a blue tint to their fur. Most bears that are light and dark brown in color are in the Rocky Mountains and west of that.

Black bears have a straight profile, a tapered, sensitive nose and a long tongue that is good for licking up insects and eating small berries.

"Hold on really tight."

"I never knew people could climb trees. Keith, what are you doing?'

At every age black bears are good tree climbers, excellent swimmers, and they can run at speeds of more than thirty-five miles (fifty-six kilometers) an hour.

Black bears will eat almost anything. Vegetation makes up the largest part of their diet. They also eat nuts, insects, fish, small mammals, birds, eggs, honey, berries, carrion, and human food. Black bears can smell much better than humans. At times they will stand on their hind legs sniffing the air and focusing their eyes on whatever is in the area. The bear's large, round ears are well developed and extremely sensitive. Normally black bears are very quiet animals. They will make woofing sounds and whimper. Cubs will cry almost like human babies. Occasionally they will growl. Staring and body movements are very much a part of their language.

"There must be something else around here to eat besides grass."

"I know you are around here."

"Give me more room."

"You stay there and fish."

"I told you to stay there and I meant it."

Black bears are sexually mature between the ages of three and four. The most active time for mating is May through June.

"I see you with another bear yesterday?"
"No, I wasn't me."

During late October or early November in the northern regions a black bear will dig a hole in the ground, usually on a hillside, then crawl into it and go to sleep. Instead of a hole, a black bear may crawl into a hollow of a tree, a cave, under some rocks or a camp, or even under a pile of leaves.

"Is anybody home?"

Cubs are born in January or early February. At this time the cubs are toothless, hairless, and cannot open their eyes for about forty days. A sow black bear could have one, two, or three, but rarely four, cubs at a time. She will keep her cubs with her until they are eighteen months old. During the first year the sow will be very protective of her little ones.

Black bears living in the wild could get to be twenty-five years old. Their primary enemies are hunters. Occasionally grizzlies will kill black bears and feed on their carcasses.

"Mom, where are you?"

"You should have been here yesterday. I met two hikers on the trail. The girl fainted and the man wet his pants."

20

"If I jump I could catch that fish. On second thought I don't think I will."

"Don't you ever scare me again, Keith."

21

"I feel sick so don't bother me with your foolishness today."

Tapeworms live as parasites in the intestines of animals and humans. They are flatworms with a segmented body. The next day there was no sign of any tapeworms coming out of this bear's body.

"Boy, it's nice to some vegetables after so much fish"

Hmmmm, which one...

"This should do me."

23

3: Spirit Bears

Spirit bears are part of the North American black bear family. These white black bears live in southeast Alaska, and on islands along the coast of British Columbia. I have also seen them more than 100 miles (160 kilometers) inland on the mainland of British Columbia. It's not unusual for a family of bears to have different colored cubs.

The fur on spirit bears is white or cream colored due to a recessive gene within the population. The white bears that I have seen have brown eyes and their claws are ivory in color. Officially named Kermode bears, Native North Americans also refer to them as "ghost" and "spirit" bears.

Black bear with two cubs (one black and one white).

"Mom, is that a bear standing on his hind legs?"

"No he thinks that he's a bear, but he's a human."

"Where are you going?"
"Nowhere. Quit following me!"

"Okay, I'll go to the pond and catch a fish."

o fish here."

25

"I'll roll in this gravel to get rid of the flies."

Then I came upon two adult spirit bears.

"Let's wrestle."
"Okay."

"Ouch! Don't be so rough."

"If you don't leave I'll stay here all night."

27

"What are you doing, hanging around with me so much?"

"Right now I want to cool off. It's too hot to be moving around much."

4: Grizzly Bears

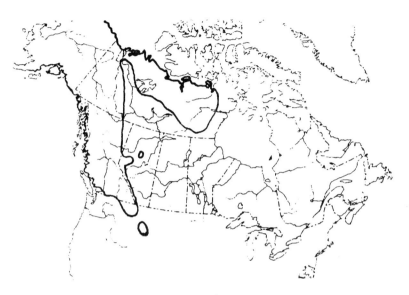

Areas where grizzly bears are commonly seen.

A grizzly bear's face is dished and flat-looking compared to a black bear's face. Their front claws can grow to be three inches (seven and a half centimeters) long and they have a pronounced hump on their shoulder. This hump is bone and muscle.

"I see you around here all the time. What do you want?"

29

On many grizzlies found inland, the end of their fur is white to silver in color.

Brown bears are grizzly bears. The reason that they are referred to as brown bears is that within fifty miles (eighty kilometers) of the coastline most of the grizzlies are light to dark brown in color. Some will be black, a few black and white, and once in a while, a silver-tipped bear can be seen near the coast.

"What kind of a bird is that?"

"I'm the prettiest bear around here."

"What do you want here?"

"What was that?"

Kodiak bears are grizzlies also. The grizzlies on Kodiak Island can get to weigh as much as 1500 pounds (680 kilograms) and grow to a size of over ten feet (three meters) in length. The biggest grizzly I have ever seen while hiking on Kodiak Island was about 1100 pounds (495 kilograms).

Grizzlies can smell far better than humans. They have good eyesight and can make out moving objects more than 150 feet (forty-five meters) away but they seem to have a more difficult time making out objects that are not moving. Quite often a grizzly will stand on its hind legs sniffing the air and trying to focus its eyes in on what's in the area. The sound of something nearby will make the bear stare in its direction or stand on its hind legs.

Though a grizzly's food consists mostly of green vegetation like grasses, clover, and roots, as well as berries, salmon, small mammals, and carrion, they will also feed on human garbage. Once seals appear in the salt water at the mouth of the rivers this is a good indication that the salmon are starting to make their way to the spawning beds. There are five kinds of Pacific salmon that spawn on the Northwest Coast. They are the chinook, pink, chum, coho, and sockeye. All five species, both male and female, die after spawning. Bears prefer catching live salmon but they will also eat the dead ones. The skin, eggs, and top part of the head are the richest and preferred portions of the salmon for the bears.

"This grass by the water tastes better."

Coho salmon.

Before the salmon arrive in the spawning beds the bears move around quite a bit in search of food.

"ot much to eat here."

"Oh no, I smell Keith. He probably wants me to go and scare some tourists with him. I'm not going."

Once the salmon arrive, the eagles appear. When bald eagles mate they stay together for life. They have their young during the spring. Immature bald eagles, less than five years old, are distinctively marked with a brown and white mottled pattern. The mature eagles have a white head and tail on a brown body. Their call sounds like a sharp screeching or "chip" sound. The eagles feed on the dead salmon and other carcasses besides catching live fish and small mammals.

"I'm going to grab you in my claws."

"You're heavier than me. I'll drag you to the shore."

With the salmon arriving in the spawning beds, more bears are coming to the creek.

"Don't be mean. Save me some."
"Go catch your own."
"Okay I will."

"Got you."

Now the eagles are feeding on salmon parts left by the bears.

"That's my fish."

"Get away, get away!"

"There is plenty here so let's try to get along."

"Cool it."

"Now what are you trying to do?"

The peak season for grizzlies to mate is May and June.

"I mean it, get out of here."

"Mom sure is worked up."

A sow grizzly will keep her cubs with her for two and a half years, and sometimes longer. At a certain point she will then chase her cubs away, sometimes up a tree, and from this time they will have to survive on their own.

"Here I come. I didn't mean to scare you Keith. I'm after a fish."

"There he goes."

"Got it. Come on over Keith and I'll give you some."

41

"Okay, okay, I'm leaving."

When bears meet usually they will stare at each other then t[...] smaller bear will leave. When this grizzly and black bear met, t[...] black bear backed away and gave the grizzly more room.

Should a big grizzly meet a smaller grizzly, a chase will usual[...] occur. Once in a while when bears meet, a fight will occur. Th[...] boar grizzly was feeding on the carcass of a smaller bear that he h[...] just killed.

"You look down stream and I'll look up here."
"Mom, are you paying attention to me?"

Mc Neil River, Alaska
with Jim Kahl

"What's that guy doing, Mom?

"This is my spot."
"That's what you think."

43

(right) "Let's wait here until they leave." Photo: Bill Kirsopp

(below) "Got ya!" Photo: Bill Kirsopp

"It's okay now but stay close to me."
*Brooks Falls, with Bill Kirsopp. Photo: Bill Kirsopp

"I'll just wait here until a fish swims by."

"Stay really close to me. There's a bear downstream."

"I don't want you to get mad at me, but would you please buzz off."

45

"Cut that out. No more foolishness."

August is the time when the huckleberries become ripe. They arc similar to blueberries.

"This is a nice change as I'm getting a little tired of eating just salmon."

"I'm going to juggle this fish. You go ahead and fish."

"I can catch a fish too. Mom, quick. Come here!"

"om, didn't you hear me. Come and see Keith trying to catch a fish in his mouth."
o, I've seen him fishing before. Everybody knows he has some screws loose. Come
re!"

"What does that mean mom — 'he has some screws loose'?"
"It means he's a nut."

49

50 *"Take off. The wolves are coming."*

5: Wolves Fishing with 🐾 Bears

"Come on up here. It smells really good."

Wolves are the biggest members of the wild dog family. They have a bushy tail, erect ears, long legs, and a wide head. They can be white, black, reddish brown, and grey in color. Wolves will travel in packs. When they howl this is their way to communicate with the rest of the pack.

"Lots of fish here."

"The bears have left a lot of fish parts."
"I'm going to get that fish from that bear."

"I'll just sneak up and grab the fish."

"Move!"
"No."

"You can put your head down and threaten me all you want. But you won't get my fish!"

"I'll get my own. There's one!"

"I missed, but there's another!"

"Got her."

"Okay, okay, cool it."

This black bear didn't want the wolf near her cubs.

"Whoo, whoo, whoo, let's get together

"I'll take this fish with me."

"I'm getting tired."

Bears and wolves rest a lot during the warm time of the day.

"I'm just going to rest my head right here."

Just before dark, and early in the morning is the coolest time of the day, and this is when the bears are active during the hot months of the summer.

As early as September many of the chum and pink salmon have already spawned and died. At this time many of the bears will eat the dead fish.

"I'll throw this salmon up in the air."

"Whoops, I missed."

"Yuck! Mom, this one doesn't taste too good. Let's give it to Keith."

"Stay away from my family. I mean it. You do too many stupid things."

During the months of October or early November, depending on the weather, a grizzly will dig a hole on a hillside. This hole will be the bear's winter home. It is built just big enough to fit. Some bears will line the bottom of their den with green vegetation.

The bears will still feed in the den area building up their body fat so that they can survive the winter sleeping period. Near a den, a sow grizzly and her two cubs were feeding on vegetation. All three bears were in good shape, having lots of fat.

A grizzly den.

"Mom, how come you're darker than us?"

"This tastes good."

A single boar continued to feed on green grasses.

Further up the mountain in the alpine zone the fall colors have arrived in the area. A sow grizzly was still nursing her two cubs. They were rolled up like a ball and looked like a big boulder.

When the snow began to fall this boar crawled into his den and fell asleep.

While asleep a bear's heart rate slows down and its energy requirements decrease considerably. This is the time when bears survive on their excess body fat. Along the coastline some bears have been known to stay awake all winter, but most bears will sleep throughout the winter until the warm weather arrives.

"Z, z, z . . ."

6: Polar Bears 🐾

Areas where polar bears are commonly seen.

The Polar Bear is largely a marine mammal, feeding along the coastal shoreline and icepack. Polar bears can weigh as much as 1500 pounds (675 kilograms). A common weight is around 600 pounds (360 kilograms).

"I'm in charge of this area, so be careful!"

"I smell something funny."

"I like you."

A polar bear's fur is pale yellow to pure white in color. The nose and part of the face are black. The skull and neck is longer than that of other bears.

The teeth of a polar bear are very sharp and they are shaped to be good for shearing meat, which forms close to 90% of their diet. When playing with other bears they open their mouths but they are careful not to bite unless fighting.

Most of their food is seals. They will spend a lot of time lying down by blowholes and openings in the ice ready for a seal to appear for a breath. When a seal sticks its head up the bear will strike it with his sharp claws or quickly bite it, then drag the seal out of the water onto the ice.

Normally polar bears will eat only a portion of the seal. It's not unusual to see an Arctic fox near a polar bear as they follow them around in order to feed on leftovers.

"I'm getting bored."

Arctic fox
"zzz..."

"You're the nicest looking bear I've ever seen."

"You probably tell that to all the bears."

"You're full of baloney, or seals, but I do like you. Give me a hug."

"I said a hug, nothing else."

Mating season occurs in April and May and polar bears are sexually mature between the ages of four and five years.

"On second thought."

"Now you've got me completely exhausted."

"zzz..."

65

A sow polar bear will have her young in December or January in their den, which is in the snow or sometimes in the earth. A mother polar bear will keep her young with her until they are two and a half years old.

Polar bears travel very long distances hunting for seals on ice flows.

It isn't unusual for polar bears to swim and travel on an iceberg. Sometimes they will be carried as far south as Newfoundland.

This is an ice flow between Labrador and Newfoundland.

7: Precautions

Both black bears and grizzlies will travel along roadsides eating the grasses, clover, wild flowers and berries that grow there. Should you stop to see a bear, do not get out of the vehicle. Bears are often accustomed to seeing cars but they are not used to encountering humans up close. When they get nervous in such a situation, they will regularly act aggressive, especially a sow with cubs.

I never stand in front of a bear or get in its way. Always give a bear the right of way.

When hiking, I avoid taking scented items along with me, such as soap, shaving lotion, and foods like fish and bacon.

Women should avoid taking powder, perfume, or scented soaps. Be especially careful during your menstrual period and stay as clean as possible because bears can smell blood from a great distance.

Bears rest a great deal during the daytime, so when hiking in thickly wooded areas I like to talk or sing. I have found that bears are more afraid of me than I am of them. I never whistle because marmots whistle in a way that is similar to humans.

It is never wise to camp near your food. You should store it 300 feet (ninety meters) away from the tent. It is a good idea to hoist food up a tree before going to bed when in a place that is populated with bears. I have been careless in the past and there have been bears that have come to visit my campsites.

It is best to use a triangle method of camping where you eat 300 feet (90 meters) away from your tent, with food storage another 300 feet (90 meters) away from the tent.

"Something smells good here."

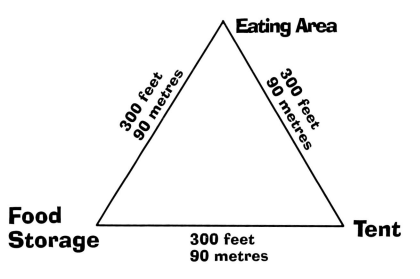

Eating Area

300 feet
90 metres

300 feet
90 metres

Food Storage

Tent

300 feet
90 metres

Drawing of food storage, eating, and tent triangle.

Burying garbage in the woods does no good, as bears can smell anything that is three feet (ninety centimeters) below the surface. If you are strong enough to carry food into the woods, then you ought to be able to take the garbage back to civilization.

"I know there is something down here"

When a grizzly comes towards me with its head down low, I know that he or she wants more room. I face the bear and talk really low, then back up slowly. Once the bear is within six feet of me I resort to yelling loudly at the bear. Should I have no bear spray with me, I drop to the ground and protect my head as much as I can. If it's a black bear I will be aggressive and swing at the bear with a stick or my hands.

Since bear spray has been invented, I always take it with me. I have used it four times. Each time the bear was less than six feet away from me when I sprayed it with the sharp stream directly in the face. Every time the bear has turned and run in the opposite direction. The spray irritates the bears eyes, nose, and skin but the effects only last about one hour. The spray does no permanent damage to the bears, as I have seen first hand. If it did I would not carry it.

Living in bear country requires extra diligence on the part of humans. Human garbage will always attract bears. Don't approach a bear feeding on garbage or anything else. The bear will be very aggressive as he or she thinks you are after the food. If you store your garbage outside, make sure that it is well sealed so that odors cannot escape. It is also a good idea not to place garbage at the end of the road until the day of collection, not the previous night.

To school children:

Often it is children who are walking to school or waiting for the bus on garbage day. If you should see a bear, back away slowly and head home. Be careful not to make sudden movements and don't go in the same direction as the bear. The fields around many schools often have a great deal of grass and dandelions that can be attractive for a bear. These foods are a delicacy for wild animals during the spring and fall of the year. It is never safe to approach or throw rocks at wildlife.

"Thanks for growing this grass for me."

"I love these dandelions."

Cattle, sheep, chickens, and other animals are potential meals for bears and wolves. Wolves and bears will try to catch dogs, cats, ponies, and horses for food. When walking in wooded areas where bears are known to live, don't let your dog run loose. Dogs will chase bears and in return bears will chase dogs, and your pet will tend to run back to you! This is a recipe for serious injury or death.

Keep dogs on leashes. Pet owners are responsible for their pets so don't let a dog loose on the way to school. Bird feeders won't just attract birds. Bears and other wildlife will also be attracted by this food. Compost heaps will also bring wildlife into your yard.

Jogging in wooded areas can also excite a bear into giving chase. Humans cannot outrun bears but groups of people will always be intimidating to excitable bears.

Black bears and grizzlies can climb trees whether the tree is heavily limbed or not. This is important to keep in mind when thinking of outwitting a bear by climbing a tree.

"I don't need any limbs to climb up or down this tree."

"Does this ever feel good! Come over and try this."

If you are camping in a trailer and it begins to rock back and forth suddenly in the night, there is a good chance that a bear is scratching itself on one of the corners of the trailer. If you shout it will likely leave. Bears love to scratch on trees, telephone poles, rocks, and houses as well.

73

Bears are not stupid. They know that we are food carriers. In good fishing areas both grizzlies and black bears will approach humans hoping for a meal of fish.

It is a good idea, when hiking in bear country, to stay at designated campsites. If there is garbage around or a lot of small animals in the area, don't camp there. It is also somewhat dangerous to camp on or near hiking trails since bears move along these like humans do. When hiking, watch out for bear signs like tracks, scratches on trees, bear tunnels, fish parts, scat, and a number of birds flying in one spot. The birds indicate a carcass and bears will sometimes remain in the area of a kill in order to protect it, so it is a good idea to circle around this area.

Bear tunnel

This is a silver-tipped grizzly protecting its caribou kill.

"I'm warning you, don't come any closer."

If you see tracks in the mud or snow, put your own track next to them and this will give you an indication of how recently the tracks were made.

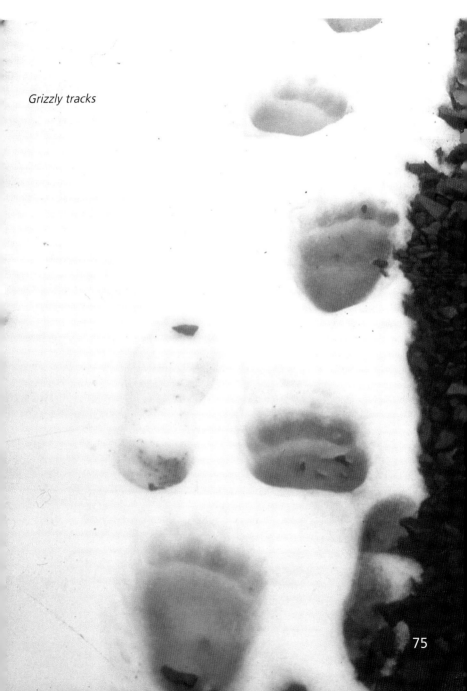

Grizzly tracks

 "These are my berries Keith. You stay there and I'll stay here.

"Keith, go ahead and run. I can catch you easily.

"Mom, I hear something"

Don't approach a cub that is alone. A sound from the cub will bring the sow to her side in a hurry.

When a bear is mad about your presence, they will put their head down low and move towards you, sometimes slowly and sometimes very quickly.

"Cool it. I'll back away."

If you are on a hunting trip it is good to know that the noise of a rifle shot may attract a bear. It is likely that if a bear is in the area he will follow you and may show up as you are carrying out or cleaning your kill.

Because I hike looking for bears and bear signs in order to study and photograph the animals, I have had a lot of exciting experiences and many close encounters. Any danger that I have found myself in was always my own fault. Should I ever be injured or killed by a bear, it will likely be my own fault, not that of the bear.

By taking some of these precautions the frequency of bear attacks or other incidents can be dramatically reduced. When we hear of incidents between bears and humans it may be a good idea to ask if the people involved have taken precautions to protect themselves. We are naturally equipped to be responsible for ourselves and our actions, while the bear is compelled by habit and instinct to be impulsive and violent. It is my belief that this can only help everyone in the short and long run of things. This is why I say, don't blame the bear.

"Those animals walking on their hind legs are humans. They are the most dangerous animals in the world. Some of them kill bears for the fun or 'sport' of it. They go around mounting our heads on walls, yet they act like we are the bad ones?"

I present a nature slide lecture program entitled "Be Bear Safe". One hour or one school period for one dollar per student, or a minimum $200 for a full day. This also includes a basketball demonstration.

I also offer a night program, "Hiking in Bear Country."

Keith Scott
66 Sunset Drive
Fredericton, N.B., Canada
E3A 1A1
Phone: 1-506-472-1825

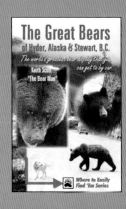